CW01582783

# LONDON NORTH

Images at and accessible to the Transport Treasury archive

200 Years of Train Travel Since 1825

ISBN 978-1-917776-18-9

First published in 2025 by Transport Treasury Publishing Ltd. 16 Highworth Close, High Wycombe, HP13 7PJ
Totem Publishing, an imprint of Transport Treasury Publishing.



www.ttpublishing.co.uk

Printed in Tarxien, Malta by Gutenberg Press Ltd.

**Front cover -** A picture that epitomises the appeal of post-war trainspotting. The young lad mesmerised by the 'living' machine before his eyes could be any one of thousands of us now rather mature railway enthusiasts, clad in his mandatory school blazer, short trousers and grey socks. For very little financial outlay he is indulging in his hobby, causing nobody any trouble and unwittingly expanding his knowledge of science, history and geography. The object of his attention is B2 Class 4-6-0 No. 61607 *Blickling* at the head of a matching set of BR Mark 1 'plum and custard' corridor coaches at Liverpool Street station awaiting departure with an express to somewhere in East Anglia, confirmed by the headlamps on the buffer beam and the line indicator showing the route as the Down Main. 61607 was one of ten locomotives re-built in 1945 by Edward Thompson from Sir Nigel Gresley's Class B17 introduced in 1928. The shedplate shows the home depot to be 31A, Cambridge which dates this picture to the late 1950's and matches the schoolboy's attire and judging by the locomotives appearance more likely to be around 1957/58 as it was withdrawn from service at the end of 1959. *(R C Riley)*

**Frontispiece -** Despite the huge variety of locomotives enjoyed by enthusiasts in North London, the Western Region seemed to offer something a little different, either the appeal of so many named locomotives or maybe that little extra of copper-capped chimneys and domes? Collett GW 'Castle' Class 4-6-0 No. 5029 *Nunney Castle* glides past Royal Oak Underground station with the Saturdays Only Up 'Torbay Express' from Kingswear in Devon, the Western Region's chocolate and cream Mark 1 coaches contrasting nicely with the red of the Hammersmith & City Line Underground services. No. 5029 was allocated to Newton Abbot shed for six months in 1959 so this may help date this view as Dick Riley provided very few clues to identify his otherwise superb collection of slides. *(R C Riley)*

**Rear cover -** A fine elevated view of Top Shed at Kings Cross (34A) with plenty to please everybody. New diesels are encroaching this steam dominated scene in the shape of a Brush Type 2 (Class 31) and a couple of 0-6-0 DE shunters (Class 08) whilst an unidentified BR 'Britannia' Pacific is being coaled under the landmark coaling stage. Some of the class were allocated to Immingham shed from late 1961 to the end of 1963 which approximates the date of the image. The ever-present coal dust and smoke would soon disappear at the end of steam and official closure of the shed occurred in June 1963 with any remaining servicing requirements passed to Hornsey, the end of an era. *(Roy Vincent)*

# INTRODUCTION

The complexity and variety of the railway system in North London cannot be adequately described in a collection of this size but an overall flavour will hopefully be represented in a generally clockwise tour, emanating from each of the termini encountered. By expanding outwards in the Down direction from each location, various points of interest will be encountered covering a period of just below 90 years.

The historical complexity arises from so many different railway companies wishing to tap into the lucrative trade offered by the country's capital, previously monopolised by stage coaches and the intense canal system. The Great Western, The Great Central, The Midland, The Great Northern, The North London, The Great Eastern and The London, Tilbury & Southend Railways all contributed in their own way to establish themselves by causing mass destruction of the already over-crowded environment to build their 'cathedrals' as near to the city centre as possible. Later of course the Metropolitan Railway caused huge disruption throughout London constructing the cut and cover tunnels later absorbed by the famous London Underground system, admired the whole world over.

North London became a huge building site over the years, much the same as now but one positive, other than providing important transport corridors, was the clearance of thousands of slum dwellings which had manifested themselves in these areas. Many changes ensued through 'Grouping' but not so much on nationalisation in 1948 when the new British Railways settled into the groove formed by their predecessors. Re-branding of course was soon evident with all locomotives acquiring an additional first digit, much to the indignation of the old school. One exception was the Western Region which retained their old cast number plates and generally steered clear of new management interference, adopting its own chosen diesel hydraulic power under the subsequent modernisation plans. Diesel electric transmission ultimately ruled but evidence of the GWR's independence can still be evidenced in the infrastructure with platforms built to accommodate Brunel's originally adopted broad gauge.

Much damage was incurred during both World Wars but without the present day restrictions and bureaucracy being in place, the railways were soon patched up to maintain supply routes for both men and materials. For a post-war 'baby boomer' brought up in the Tottenham area, the opportunities for a job were excellent but wages on the run-down railway system were none too appealing, neither were the conditions of employment. A railway enthusiast however was spoilt for choice with so many centres to visit, usually using the coveted 'Red Rover' bus tickets to get around. Knowledge of the geography of the area increased significantly by studying freely available bus maps whilst travel by train or the Underground was excluded in those pre-Travel Card days. Clutching well-thumbed Ian Allan abc volumes, hoards of young boys would head towards the busiest termini such as Paddington, Kings Cross or Liverpool Street, sometimes having to spend hard-earned pocket money on platform tickets. When this ran out, most would have favourite view points where they would congregate near their homes such as Hornsey station footbridge or the Hogs Back elevated footpath at Harringay West.

Later on as confidence increased trips might be made to those forbidden havens of power, the engine sheds. Favourites naturally were the two adjacent establishments of Willesden and Old Oak Common where the staff usually turned a blind eye to the regular trespassing. Of course the end of steam in 1968, earlier in London, diluted the interest but after passing through adolescence and other obvious attractions, many resurrected their interest and we must all be very grateful to the band of photographers who could somehow afford the film and who continued their interest right the way through to the present day. There is still much to admire on today's railway, lines such as the London Overground and the Docklands Light Railway reinvigorating abandoned tracks, once considered lost forever. Thankfully the Transport Treasury has accumulated much of this precious material to enable us to enjoy the nostalgia and extend our interest into the present day and celebrate the 200th anniversary of the first public railway.

Kensington Olympia station on the West London Extension Railway is a convenient location to commence our study of North London being on the major link for main services travelling to and from the Southern Region. Opened as Kensington Addison Road in 1868, the London & South Western Railway as well as the London & North Western and Great Western Railways all operated services along with the Hammersmith & City Railway on a none too regular basis. Operations were seriously curtailed during the war and by the time the station had been re-named Kensington (Olympia) after the adjacent exhibition centre in 1946, the only regular service was a rush-hour shuttle service to Clapham Junction. Provided purely for the benefit of Post Office Savings Bank HQ workers in Blythe Road, these trains were not advertised in public timetables as they were initially under restrictions of the Official Secrets Act. Popularly referred to as the 'Kenny Belle', the empty stock for the 5.6pm to Clapham Junction is arriving on 22 March 1961 behind Bricklayers Arms-based BR Standard Class 2 2-6-2T No. 84022. Shortly it would be transferred to Exmouth Junction and Llandudno Junction for short sojourns out on the network before semi-retirement after overhaul as Crewe Works shunter, where 2 years later its shamefully wasted career of just under 7½ years was terminated. *(Terry Gough)*

Apart from being a regular location for inter-regional services changing locomotives, Kensington Olympia also played a significant role in providing London with milk. Trains arriving from the GWR and later BR Western Region would be divided in the station with the special tanks being shunted into the local Express Dairy creamery and bottling plant sidings. On 30 May 1964 Hawksworth GWR 0-6-0PT No. 9418 is busy handing over the wagons to an unidentified classmate, doubtless a fellow resident of Old Oak Common. Inevitably by the end of the decade all of this once lucrative business had passed over to road transport. From 1966 BR (Motorail) operated a series of daytime trains from its dedicated terminal at Olympia to Perth and subsequently Carmarthen, Birmingham and Inverness as well as an overnight service to Carlisle. With various existing connections to all over the UK, Olympia was the ideal hub for these operations but with the ever-expanding motorway network and the much improved quality and reliability of the motor car these trains ceased from 1981. 'Kenny Belle' services ended around 1986 but ironically these days London Overground and Southern operate a very regular public service over the route of the 'Ghost Train'. *(Terry Gough)*

On Sunday 22 September 1957 the Stephenson Locomotive Society ran the 'Hammersmith & City and Inner Circle Rail Tour' over London Underground metals which is seen here on arrival at the Hammersmith branch terminus behind Metropolitan-Vickers 600V DC Bo-Bo No. 16 *Oliver Goldsmith*, built in 1923 and surviving until electrification was extended to Amersham. These locomotives dominated the Metropolitan Railway outer suburban lines between Baker Street and latterly Rickmansworth where steam motive power took over as far as Amersham and Chesham. On this occasion however, steam was attached to the rear of the train, just discernable at the end of the 5-coach train, in the form of Class E 0-4-4T No. L46 (ex-MET 77), built at Neasden Works in 1896. This veteran survived until 1963/64 on goods transfer and engineering trains around the Neasden area having previously been a regular on Chesham branch trains and out as far as Verney Junction. Sister locomotive No. L44 has subsequently been restored under its earlier guise of Metropolitan Railway No. 1, albeit currently awaiting a boiler overhaul. Along with electric Bo-Bo No. 12 *Sarah Siddons* and the Bluebell Railway's rake of restored Metropolitan Railway coaching stock, a re-creation of this scene remains an enticing possibility despite ever-increasing Health & Safety restrictions. *(Frank Goudie)*

A scene that must have been recorded countless times at Paddington terminus with a representative of Collett's GW masterpiece 'King' Class 4-6-0 on arrival at platform 8 with an Up express and passengers (or customers in modern day parlance) scuttling towards the exit. This is the summer of 1961 and No. 6011 *King James I*, being a Stafford Road engine, has likely had charge of this train since coupling up at Wolverhampton Low Level and will shortly reverse post haste to Old Oak Common or maybe Ranelagh Bridge for servicing. On Platform 6 the new brigade is threatening with one of the Metropolitan Cammell de-luxe diesel sets forming either the 'Bristol Pullman', the 'Birmingham Pullman' or possibly the newly introduced 'South Wales Pullman'. It would not be until 1976 that High Speed Trains started operating regularly out of Paddington virtually eliminating the need for a change of locomotive with these iconic Class 43 InterCity 125 trains surviving until 2019. *(Barry Richardson)*

The BR Class 50 Type 4 DE Co-Co has always been very popular with the modern traction enthusiast and around 1980, when the type were about to be ousted from the most demanding services by the HSTs visible in the background, No. 50032 *Courageous* is awaiting departure from Paddington with a Down main line express having replaced earlier hydraulic types. Initially on a ten year lease from English Electric, the 'Hoovers' as they were nicknamed were initially employed on the non-electrified section of the West Coast main line, often double-headed and numbered in the range D400 to D449. Acquiring the names of battleships in the tradition of their Class 42 'Warship' predecessors, these 100mph rated locomotives were relegated to operating services to Worcester and Hereford via Oxford by the successful HSTs and later moving across to the Southern Region to work the Waterloo-Exeter services before their retirement with second generation DMUs taking on the task. *(Arthur Turner)*

The Great Western Railway built 38 streamlined railcars between 1932 and 1942, incorporating a high speed diesel engine manufactured by A.E.C. and designed generally to operate singly or supplemented by a single trailer or as two power cars with a trailer in between. These fast and efficient trains were ideal for passenger use on short to medium journeys but also operated the occasional express service on some routes. Two representatives were built as parcels units although other railcars were temporarily adapted during the war for emergency use. W34W was one of the last to be built in 1941 at Swindon Works and is observed in red livery departing from Paddington parcels depot hauling a single van past Ranelagh Bridge loco servicing point with Down 'Express Parcels' service. The Up goods yard looks characteristically busy in this view dated as between 1958 and 1959 and at least three light engines can be seen in the throat of the terminus, almost certainly engaged on empty coaching stock duties. The two young boys (far left), suitably clad in school blazers and caps, have a great viewpoint on Royal Oak, but seem somewhat non-plussed by this intrusion, much preferring no doubt fine, shining named 4-6-0s which still prevailed at the time. *(R C Riley)*

Simply oozing power at Ranelagh Bridge Loco on the same day in the late 1950s are two of the Great Western's giants of steam, a 'King' and a 'Castle'. The parallel boiler of the 8P power-rated 'King' is suitably cleaned and polished, as expected for a top link locomotive but regrettably the identity of the 84A Wolverhampton Stafford Road allocated locomotive cannot be confidently verified. Worcester-based 7P-rated 'Castle' Class No. 4088 *Dartmouth Castle* proudly displays its shining number and name plates and clearly has steam to spare, not that the residents of the adjacent tenement blocks would share the joy. Both locos are ready to reverse down to the terminus at Paddington a short distance away but tantalisingly just too far for most platform-end spotters who would have to nip out to Royal Oak by Underground if time was short. Just discernable under the far bridge is one of the batch of Collett's GW Class 5700 0-6-0PT fitted with condensing apparatus for use over the LT Metropolitan Line and numbered within the range 9700-10. *(R C Riley)*

Summer sunshine through the fanlights of the roof gives sufficient light to identify a selection of 4-6-0 locomotives crowded around one of the turntables inside Old Oak Common shed. The LMS 'Black Five' interloper is No. 44862 allocated at this time to neighbouring Willesden (1A) shed and seemingly guarded by Hawksworth Modified Hall No. 6961 *Stedham Hall* from Southall, Hall No. 5933 *Kingsway Hall* from Oxford, Grange No. 6813 *Eastbury Grange* from Worcester and resident Castle No. 7032 *Denbigh Castle*, the latter three all designed by Collett. No. 7032 seen here in steam, was withdrawn in late September whilst another view taken on the same day features new Brush Type 4 No. D1733 in its experimental XP64 blue livery which was not delivered until mid-June, thus narrowing down the date to summer 1964. *(R C Riley)*

No collection such as this should omit probably the most successful class of diesel locomotive to grace the lines of North London, the BR Class 47 Brush Type 4 Co-Co, totalling 512 in number and built between 1962 and 1968 at either Brush Traction (Loughborough) or BR Crewe Works. Survivors can still be seen operating in North London mainly on special workings, with a number based at Southall MPD under the banner of West Coast Railways though sadly not in the appealing two-tone green livery in which they were turned out. The livery of *Sir Daniel Gooch* seen here at Old Oak Common in the late 1980's gives no cause for complaint, similar GWR Brunswick Green paint with cast number and name plates being applied to sister locomotives Nos. 47079, 47484 and 47500 respectively named *George Jackson Churchward*, *Isambard Kingdom Brunel* and *Great Western* to celebrate GW150 in 1985. No. 47628, originally D1663 then 47078, lost its individuality in 1989 when it emerged from Doncaster Works following overhaul in 'Mainline' general purpose livery before withdrawal in 1997. *(Arthur Turner)*

Churchward GW Class 4700 7F 2-8-0s were relatively rarely seen in daylight hauling the express freight for which they were introduced in 1919 but on a June day in 1962, Southall's No. 4707 is powering through West Ealing on the Down Goods line with a substantial van train. A small class of just nine, these reliable locomotives were designed following the First World War to handle express freights operating between London and Plymouth, Bristol and the West Midlands. Such duties were normally pathed during the quieter overnight period from whence they acquired the nickname 'Night Owls'. However, from 1957 onwards, they proved to be very successful handling heavy relief holiday trains to the West Country, earning themselves a re-paint from the previous black livery into BR Brunswick Green. Most enthusiasts though would have noted them simmering quietly on shed, particularly Old Oak Common, awaiting their call into action. *(Barry Richardson)*

Probably on the same day, one of Collett's Class 5700 0-6-0 pannier tanks, No. 9704 fitted with condensing apparatus for use through the Metropolitan Railway tunnels to Smithfield, takes the Greenford branch at West Ealing with a milk empties and pulling away from the 'milk dock' on the north side of the Up Slow. This line formed a triangular junction with the GW/GC Joint Line at Greenford and a shuttle service still forms a connection with London Underground's Central Line whilst the so-called North Main Line is currently severed as HS2 construction work proceeds. BR Class 165 Networker Turbo units nowadays operate this often unreliable service whilst historically First Generation Class 121 single or double diesel units have been rostered as well as GWR railcars but most lamented are the push-pull steam services provided by a Collett Class 1400 0-4-2T and auto-coach. *(Barry Richardson)*

Approaching on the Up relief around the same time is one of Hawksworth's GW 1945 design 'County' Class 4-6-0s No. 1015 *County of Gloucester* with a lengthy parcels train. These 6MT-rated engines were the result in a way of the designer being prevented from producing a Pacific locomotive due to war-time restrictions. Incorporating several innovative ideas and leaning heavily towards Stanier's Class 8F, then being built at Swindon, the instantly recognisable feature was the continuous splashers over the driving wheels giving rise to their straight nameplates. Unfortunately none survived long enough to be considered for preservation although the Great Western Society are undertaking a reconstruction using major components donated from other locos salvaged from Dai Woodham's scrapyard in Barry, South Wales. In recognition of the late and canny businessman and acknowledging donations of parts from the local authority, this locomotive will hopefully soon be resurrected in the form of No. 1014 *County of Glamorgan*. *(Barry Richardson)*

Although tragically short-lived, BR's Class 35 'Hymek' Diesel Hydraulic B-Bs were surprisingly popular with enthusiasts, possibly because of their departure from more familiar appearances. 101 of the lightweight Type 3s were built primarily for secondary passenger and freight duties but often called upon for more onerous duties to which they weren't suited. The name derives from the Mekydro transmission used by builders Beyer Peacock but their main downfall was BR ultimately classifying hydraulic transmission as non-standard, a fate which also befell the 'Warships' and 'Westerns'. None of the class received names nor carried the Class 35 numbers and passing West Ealing is No. D7024 heading a Down semi-fast adorned with the original Brunswick green with lime green lower stripe livery with small warning panel. Cardiff Canton-based and only 3 months old, this locomotive would be withdrawn from service on 1 January 1972, 2½ months under 10 years old! *(Barry Richardson)*

Passing Southall MPD (81C) in the Spring of 1962 is Hawksworth GW 'Modified Hall' Class 5MT 4-6-0 No. 6973 *Bricklehampton Hall* heading a Down special working. This was part of Old Oak Common's stud at the time and it would move on in due course to see out its career of under 16 years at Cardiff East Dock, Neath (Court Sart) and Severn Tunnel Junction before withdrawal from Bristol Barrow Road in August 1965. The large goods shed on the left is now devoid of track but survives in industrial use whilst part of the engine sheds on the right now house the rolling stock and locomotives of West Coast Railways as a sub-depot of their headquarters in Carnforth. *(Barry Richardson)*

A few years earlier on 26 April 1952, the through branch service to Uxbridge Vine Street is seen departing from West Drayton at 2.12pm, originating as the 1.37pm Saturdays only from Paddington. The totem attached to the lighting standard correctly indicates the station name as West Drayton & Yiewsley, although in 1974 it reverted to its original name of West Drayton. Southall-based Collett GW 2-6-2T No. 6125 will shortly bear right as signalled across the Up Slow to take the branch to Uxbridge with one intermediate stop at Cowley. West Drayton retains five platforms to this day, the broad-gauge spaced platforms 1 and 2 on the main line rarely used apart from when engineering works are being carried out to the Up and Down relief lines. Platforms 3 and 4 are regularly used by Elizabeth Line services to and from Reading whilst Platform 5 on the Up goods line never sees passenger use but serves freight trains accessing the branch line to Colnbrook on the ex-Staines West branch which closed to passengers in 1965. *(J C Flemons)*

The complex history of the railways of Uxbridge is well documented but suffice to say the Great Western Railway never achieved its full ambition of putting this desirable suburb on the main line map. Two short branches were constructed, one off the GW/GC Joint Line east of Denham terminating at Uxbridge High Street and another off the main line to Reading at West Drayton terminating at Uxbridge Vine Street. Any intentions of routing the main lines through the town had been strongly opposed by those with interests in the well-established London to Oxford coach route and various schemes to loop the two branches together never came to fruition despite the land corridor being safe-guarded. The Metropolitan Railway subsequently introduced its electric service in 1904 to seek out custom from the rapidly increasing suburban development , competing with the tram route to Shepherds Bush which opened in 1901. Unsurprisingly, the High Street branch closed to passengers in 1939 whilst Vine Street, depicted here, survived until 10 September 1962. Frequent London Underground services cater for most needs these days, with both Metropolitan and Piccadilly Line trains operating from the combined station. This image taken during the period 1956-1958 captures Collett GW 1400 Class 1P 0-4-2T No.1436 arriving with the push-pull shuttle from West Drayton whilst one of his 6100 Class 2-6-2T locomotives, No. 6128, waits to depart with a through train to Paddington, both engines being based at Southall. *(A E Bennett)*

Probably a post-morning rush hour scene, this image, taken sometime during 1960/61 going by the dates both locomotives were allocated to Neasden at the London end of the ex-Great Central line, perfectly captures the deserted and neglected atmosphere which one normally associated with Marylebone. Whilst these two Fairburn LMS 2-6-4Ts were on suburban duties, the subsequent deliberate run-down of the main line appeared to spell the end in the 1980s for this partly hidden terminus, which unlike Kings Cross, St. Pancras and Euston, is some way removed from the connecting Euston Road. However, a reprieve was to follow with regard to increasing commuter traffic from Aylesbury and High Wycombe lines and, under Privatisation, the number of platforms increased and additional through services were introduced to Birmingham Snow Hill. The original main line had by now of course been lifted and the route severed in many places, a situation much lamented at the moment by proponents of HS2. *(Roy Hobbs)*

Marylebone station was never the most enticing of all the London termini due mainly to the sparsity of its services. Towards the end of steam however, choices were more limited and thankfully a handful of photographers forged some interesting results despite the appalling condition, external and often internal, of the locomotives working out their final days. Banbury shed, once a Great Western establishment, had its allegiance and code changed from 84C to 2D on transfer to the London Midland Region and provided motive power for the surviving skeleton service to Nottingham until the Great Central line was closed south of Rugby on 5 September 1966. 'Black Five' No. 45331 was no stranger to the London area though having operated from Bletchley since 1955 with a short interlude at Willesden. On 31 May 1966 it is seen being turned at Marylebone loco servicing area, being retained until the cessation of steam from the station. This wasn't quite the end for this workhorse as it was spared the final ignominy of rundown by being transferred to Springs Branch in Wigan to extend its life for another 18 months. *(David Idle)*

Between 1961 and 1992 outer suburban trains from Marylebone were operated by BR Class 115 Derby 4-car DMUs serving destinations as far out as Aylesbury, High Wycombe and Banbury. Power car No. 51655 heads a service into the London terminus bearing its later blue and white livery, having originally been turned out in the standard BR DMU livery of lined out green complete with 'cats whiskers' followed by the white/blue stripe applied in late 1978. The full yellow end contrasts nicely with the red of the passing 'Routemaster' bus and the well-framed home semaphore signals. Identification of the parked up learner's road vehicle is best left to experts in that field! *(Arthur Turner)*

Neasden shed (34E later 14D) was always the place to find something unusual particularly on days when big football or rugby matches were being played at nearby Wembley Stadium. On this occasion though, Parker GC Class N5/2 2MT 0-6-2T No. 69319 is a bona fide resident having very recently moved south from Langwith Junction. A total of 131 of the very successful locomotive were built between 1891 and 1901 primarily for specialised shunting and trip freight duties but later passenger workings in the Manchester area. No. 69319 clearly in steam on 10 February 1957 entered service in March 1899 and was known to have operated the Chesham push-pull shuttle until the veteran was withdrawn on 12 December 1959. *(Nick Nicholson)*

At rest outside the London Underground depot at Neasden in the late 1950s is a spotless No. L.50, an 0-6-2T F Class locomotive built by the Yorkshire Engine Company in 1901 for the Metropolitan Railway where it received the number MET91. Intended for haulage of freight trains over the main line, all four of the class were withdrawn between 1957 and 1962. Sister locomotive No. L.52 was side-lined for preservation by the Met Tank Appeal Fund but on discovery of a cracked mainframe, E Class No. 1 (LT No. L.44) was chosen instead, as the F was very similar in design to the Class E and now resides at the Buckinghamshire Railway Centre at Quainton Road. *(Roy Vincent)*

Neasden depot was established in 1882 as a locomotive and coaching plant for the Metropolitan Railway but from 1898 concentrated on servicing both steam and electric locomotives. Currently the facility concentrates on servicing stock which operate on all the London Underground sub-surface lines as well as stabling some stock for the Jubilee Line. In this view are examples of 1938 Stock together with D Stock and soon to be phased out T27 Stock. The distinctive chimneys of Neasden Power Station, constructed by the Metropolitan Railway in 1904, can be seen in the background, sharing the generation of power for the whole LPTB network with Greenwich and Lots Road Power stations. Coal was brought in by its own steam locomotives until the LNWR took on this responsibility in 1935 and then continued by British Railways. Neasden Power Station closed in 1968 and was demolished shortly afterwards. *(Roy Vincent)*

A classic view of a Stanier Class 5MT 4-6-0 running light past Neasden South Junction box that typifies North London. An everyday locomotive based at nearby Willesden shed, as described on the buffer beam and verified by the 1A shedplate, No. 44862 was one of 842 built between 1934 and 1951, some surviving until the very end of steam on BR. Both Willesden and Cricklewood had significant numbers in their stock and they could be found anywhere on the London Midland as well as the Eastern, North Eastern, Western and Scottish Regions whilst also running excursions onto the Southern. No. 44862 is on the Down Main towards High Wycombe on 27 March 1965 whilst the single upper quadrant semaphore home signal controls movements joining the Up Main over the former Metropolitan Railway lines into London. The home and distant bracketed signal protects the route curving round behind the ex-Great Central Railway Neasden South Junction box and heading via Acton and Gunnersbury and over Kew Bridge onto Southern Region tracks. The signal box closed in 1989 and was dismantled to be later reconstructed at the Nottingham Transport Heritage Centre, Ruddington. *(David Idle)*

Wembley Hill station was situated on a freight route opened by the Great Central Railway between Northolt and Neasden Junction in 1905 with passenger services inaugurated the following year with the opening of stations at Sudbury & Harrow Road, South Harrow and Wembley Hill (re-named Wembley Complex in 1978 and Wembley Stadium in 1987). Confusion often arises since the LNER opened a station called The Exhibition Station (Wembley) in 1923 for the British Empire Exhibition. This was situated on a loop off the Chiltern Main Line between Wembley Hill and Neasden Junction with both connections facing London to facilitate a frequent service without the need of reversal. This station closed in September 1969, last being used for the 1968 FA Cup Final between Everton and West Bromwich Albion. Baggies fans will need no reminder that their team won 1-0 after extra time with the late Jeff Astle managing to score in every round. The date here is 11 May 1963 with the occasion being the final of the Rugby League Challenge Cup between Wakefield Trinity and Wigan. Fans alighting from the 1X60 excursion from Wakefield will later be celebrating a 25-10 victory with star player Neil Fox converting five goals. Jubilee Class 4-6-0 No. 45568 *Western Australia* was a Holbeck-based locomotive at the time. *(Alec Swain)*

Stanmore station was opened in December 1932 as the terminus of a branch spurring off the Metropolitan Railway main line north-west of Wembley Park with intermediate stations at Kingsbury, Queensbury and Canons Park. The stabled T27 rolling stock on 10 October 1937 is marked as 'empty stock' but will not be gracing these platforms much longer. Following incorporation by London Transport, operation transferred to the Bakerloo Line which thereafter employed its own Production Stock until replaced by Standard 1938 stock from 1971 and eventually becoming part of the Jubilee Line in 1979. Nearby the London & North Western Railway operated a single line branch from Harrow & Wealdstone to Stanmore (later Stanmore Village under BR) between 1870 and 1964 but could not compete with the service provided by the new electric railway, mainly due to its north-facing junction. In 1952 passenger services were cut back to the intermediate station of Belmont and the line finally fell victim to a certain Dr. Beeching in October 1964. *(Milepost)*

Rail Express Parcels
Euston Depot
WATFORD
81

**Left:** Departing from Euston in the early 80s is a BR Class 501 3-car suburban unit, the Standard design and of little interest to railway enthusiasts by the time they received their BR Rail Blue livery. Notwithstanding this remark, two individuals can be seen observing activities from the platform ends, once crowded with 'gricers' hoping for something rare to turn up from the North-West or even Scotland. Most would have only considered these units useful for transport out to Willesden Junction to 'bash' the sheds at 1A and 81A, gaining en route, through the barred windows, tantalising glimpses of Camden shed (1B) provided an empty stock working didn't block the view! *(Arthur Turner)*

**Right Top:** In 1997 Virgin Trains were awarded the franchise to operate InterCity West Coast services and placed an order with Alstom to produce 57 high speed tilting 'Pendolino' Class 390 units following BR's unsuccessful trials with the Class 370 Advanced Passenger Train. With a maximum speed of 125 mph due to a lack of signal upgrading, these 9/11-coach units have provided the backbone of expresses on this route since 2001 despite the operator changing to Avanti West Coast. 11-car unit No. 390151, originally named *Virgin Ambassador* is clearly waving the original operator's flag and emblazoned 'Business is Great' as displayed on 28 February 2015 on its arrival with an Up service at the efficient but characterless Euston station, as re-built in 1968. *(Author)*

**Right Bottom:** The old brigade in the shape of the Class 90 still managed the occasional appearance in charge of the overnight Caledonian Sleeper services seen on the same day with EWS No. 90024 painted in First ScotRail colours. Late that year the contract passed to GB Railfreight who opted to use the more modern Class 92 electric locomotives. *(Author)*

Sir William Stanier learned his trade under the GWR at Swindon Works, becoming the Chief Mechanical Engineer of the London, Midland & Scottish Railway in 1932. It was only natural to find several similarities in his first design of an express passenger locomotive intended to at least match the performance of their greatest competitor, the London & North Eastern Railway. The 'Princess Royal' Class 8P 4-6-2s, thirteen in number were clearly based on GWR No. 111 *The Great Bear* with other components not far removed from those on the GWR 'King' Class 4-6-0. Each of the locomotives were named after a princess and the class after Mary, Princess Royal as she was the Colonel-in-Chief of the Royal Scots. Most staff referred to them however as 'Lizzies' after Princess Elizabeth who later became our late Queen. No. 46200 *The Princess Royal* is seen here in Camden shed yard on 3 June 1962, re-fuelled after working the return RCTS/SLS 'Aberdeen Flyer Rail Tour' from Carlisle to Euston. Little work remained available for this splendidly turned out locomotive and in November it was withdrawn from service and broken up in September 1964. *(R C Riley)*

Still on top link work at the same time was Stanier's improved and enlarged version of the 'Princess Royal' design, the 'Coronation' Class Pacific, again rated as 8P but proven to be probably the most powerful steam locomotive ever to run on the network. Ten of the original batch built from 1937 were turned out in streamlined form, five of which were dedicated to haul the prestigious 'Coronation Scot' express. The casing was subsequently removed but they still dominated express and heavy workings until the gradual takeover by the English Electric Type 4 DE 1Co-Co1 from 1962. An unidentified member of this later BR Class 40 sits alongside No. 46220 *Coronation* at Camden MPD in June 1962, not the first built but holder of the speed record at 114mph in 1937 until usurped by Sir Nigel Gresley's Class A4 No. 4468 *Mallard* in 1938. The Coronations carried various liveries in their careers including the BR lined out Brunswick Green carried here. A Carlisle Upperby locomotive at the time, this was its final home with withdrawal and rapid scrapping occurring in April 1963. *(R C Riley)*

A
STONEBRIDGE PK
BAKERLOO LIN
2011

**Left:** Queens Park London Underground station is situated in West Kilburn, opened originally by the LNWR in 1879. As part of the Bakerloo Line trains once ran to Watford Junction but nowadays these services are curtailed at Harrow & Wealdstone, those beyond catered for by London Overground Lioness Line services. In this undated shot the once familiar red livery of London buses and tubes clearly defines the profile of 1938 stock, probably more recognisable to younger readers as the trains that operated from Ryde Pierhead on the Isle of Wight system. *(Arthur Turner)*

**Right Top:** A regular visitor to Willesden shed, No. 48898 started life late in the nineteenth century as a London and North Western Railway A Class 3-cylinder compound 0-8-0 designed by Francis Webb. 110 were built but not without complications so Webb's successor George Whale and later Charles John Bowen-Cooke set about re-building these as Class D with simple expansion engines, some such as this Bletchley-based engine subsequently acquiring a G2 boiler and Belpaire firebox. Despite the influx of more modern Stanier 2-8-0s and BR 9F 2-10-0s, these veterans were still a sight to behold plodding along the West Coast main line. 48898's days were sadly numbered though as in March 1962, shortly after this picture was taken, it was transferred to Buxton but soon withdrawn at the end of June after a service life of over 59 years. *(Barry Richardson)*

**Right Bottom:** British Railways prototype Class D16/2 1Co-Co1 diesels were actually designed for the Southern Railway by none other than Oliver Bulleid which may be apparent from the uninspiring profile. With English Electric providing diesel engines and transmission, No. 10203 was the third and final example to arrive, from Brighton Works, four years after Nos. 10201 and 10202 in July 1954. Allocated initially to Nine Elms, it was transferred to the London Midland a year later and shared its time between Camden and Willesden sheds (seen here in Winter 1961/62) and more significantly Derby Works. With its unacceptable time out of service and the ever-increasing cost of repairs it was consigned to store at Derby in October 1962, officially withdrawn on 7 December 1963 and eventually dismantled for scrap at Cashmore's, Great Bridge in the West Midlands. During the author's frequent visits to Willesden, all three of the class together with the older LMS prototype pair of Co-Cos Nos. 10000 and 10001, had often been shunted well out of the way of daily activities. Notwithstanding the above, the bogie and power train formed a sound basis for the early representatives of BR Class 40 Type 4, albeit in the more traditional English Electric body shape including American-style nosing. *(Barry Richardson)*

A classic locomotive in a classic location. Stanier LMS 'Royal Scot' Class 7P 4-6-0 No. 46115 *Scots Guardsman* awaits its next turn of duty outside the running shed of Willesden MPD with coal piled high in the tender. These powerful locomotives were introduced in 1927 by Sir Henry Fowler, initially with parallel boilers, to successfully eliminate the requirement for double-heading on the West Coast main line. Between 1943 and 1945 the class was re-built with tapered boilers and angled top deflectors to solve the problem of drifting smoke which much enhanced their appearance. *Scots Guardsman* was allocated to Longsight shed during this spell and will likely soon be heading back to Manchester with a Down express. Fortunately this locomotive together with No. 46100 *Royal Scot* has been actively preserved. This appears to be a very busy day, sometime between the summers of 1961 and 1964, with all locos in view primed and ready to leave, the fireman of the partly hidden 'Jubilee' Class 4-6-0 No 45676 *Codrington* seemingly having just placed a few more shovelfuls of coal into the firebox. *(R C Riley)*

With early electrification of the West Coast main line, high speed fixed unit trains were very much a thing of the future. A series of AC Bo-Bo locomotive prototypes from different manufacturers were being trialled during the early 1960s from which the AL6 or BR Class 86 was developed and 100 subsequently built between 1965 and 1966. No. 86229 previously No. E3119 is viewed inside Willesden TMD rubbing shoulders with one of the unsuccessful AL2 candidates produced by AEI No. 82006 which soldiered on until 1983 on secondary duties. The traction maintenance depot was gradually developed after the closure in 1965 of the steam depot on the opposite side of the main line. The Class 86 was itself removed from the most demanding services by the Class 87 higher powered locos but continued hauling regular passenger trains from Liverpool Street to Norwich until 2005 when they were replaced by the Class 90. Their use on Freightliner workings continued until 2021 and three are currently preserved, No. 86259 *Peter Pan/Les Ross* still in regular use on main line excursions. *(Arthur Turner)*

Willesden Junction is currently the unconnected intersection of two London Overground Lines. On 3 January 2015 a Class 378 Capitalstar 5-car unit is passing with an eastbound service towards Stratford. These services have now been re-branded as the Mildmay Line whilst passing beneath the bridge is the Lioness Line, operating between Euston and Watford Junction and sharing tracks with the Bakerloo Line as far as Harrow & Wealdstone, evidenced by the presence of the fourth rail. Beyond the rear of the train is the West Coast main line which until 1962 had Low Level platforms, swept away during electrification works. The Class 378 is part of the Bombardier 'Electrostar' family specifically designed for London Overground with longitudinal seating, similar to Underground trains, to reduce overcrowding. *(Author)*

A Down Bedford line suburban service ensures home-going City commuters are provided with a liberal coating of soot as it enters Aldersgate & Barbican station on 13 May 1960. Opening in 1865 as Aldersgate Street, the name was shortened in 1910 to plain Aldersgate then re-named as Aldersgate & Barbican in 1924. During World War II much damage was inflicted during the Blitz, disrupting services particularly in December 1940. A final name change occurred in December 1968 to Barbican. Steam hauled services were replaced by diesel multiple units and Brush Type 2 (BR Class 31) A1A-A1As hauling steam age stock until electrification on the Great Northern suburban lines in 1976. Midland Line electric services started operating in 1982 with trains from Bedford forming part of the Thameslink network although these platforms closed again in 2009 under the 'Thameslink Programme' to allow for Farringdon's platforms being extended across the Moorgate branch. This has resulted in the current Barbican station only being served by Underground services but with direct access to Farringdon for Elizabeth Line services which, prior to the eventual opening were commonly referred to as Crossrail. *(James Harrold)*

Moorgate was originally a terminus opened by the Metropolitan Railway in 1865 as Moorgate Street. Following increased traffic generated by the Great Northern Railway, the tracks between Kings Cross and Moorgate were increased to four, hence the City Widened Lines title. GNR Up suburban services through Kings Cross connected via tunnels at York Road with Down services returning via Hotel Curve on the opposite west side of the station. Similarly, Midland Railway suburban services to the City connected to the City Widened Lines via a junction at Kentish Town. On 13 May 1960, Fowler LMS Class 3P 2-6-2T No. 40038 is awaiting departure from Moorgate with a Down Bedford line suburban service formed of BR standard compartment stock. Sometimes referred to as Sir Henry's most unsuccessful design, some soldiered on for 30 years, the class forming the basis for Stanier's subsequent tapered boiler 2-6-2T. 40038 was one of 19 of the class of 70 fitted with condensing apparatus to operate through the Widened Lines which is clearly seen here together with the Westinghouse brake pump. The 14A shedplate affixed to the smokebox door confirms its allocation to Cricklewood at the time, sharing most of its career between there and Kentish Town. *(James Harrold)*

From the mid-1960s through to much of the 1980s there was very little variation in the main line motive power to be seen at St. Pancras with Class 45 'Peak' locomotives and later HST Class 43 units monopolising procedures. Construction of Sulzer-engined 1Co-Co1 Type 4s was split between Derby and Crewe Works from 1960 to 1962 and based on the pilot scheme 'Peaks' from which they derived their names. Reposing beneath the imposing arched roof of the Midland Railway's showpiece station is No. 45118 *The Royal Artilleryman* based in the Nottingham division from where it has likely delivered this service. 45118 was re-numbered from D67 on 12 December 1973 and withdrawn on 8 May 1987, whereas classmate No. 45140 on the adjacent platform was re-numbered from D102 on 30 October 1974 but withdrawn later. 18 July 1981. *(Arthur Turner)*

Rush hour at Kentish Town on 7 September 1959 and St. Alban's-based Fowler LMS 3P 2-6-2T No. 40024, suitably fitted with condensing apparatus has emerged from the Widened Lines tunnels with the 5.36pm Moorgate to Harpenden Central commuter service. Less celebrated and almost hidden alongside is a classmate undoubtedly based in the local shed awaiting departure with a service around the spur onto the Tottenham & Hampstead Junction line with a Barking service. These days the spur has disappeared and services diverted to Gospel Oak and extended eastwards to Barking Riverside. Prior to the removal, boat trains from St. Pancras to Tilbury and seaside excursions to Southend and Clacton would regularly travel this way. *(Leslie Freeman)*

Kentish Town shed was always a gloomy, smoke-filled establishment but could always be relied upon to produce the odd visitor or two from further afield to complement its normal stud of condensing tanks and primarily freight locomotives. Nottingham-based Jubilee 4-6-0 No. 45620 *North Borneo* has just had its tender replenished and will shortly be turned ready to back down through Camden Road Tunnels to St. Pancras. Crammed in amongst the obvious industrial background can be seen traces of residential properties whose occupants would have become seasoned to the smoke and noise emanating from the railway property twenty four hours a day. *(Transport Treasury)*

Contending for the title of most unattractive member of the British Railways Pilot Scheme for diesel locomotives must be the Class 28 DE Type 2 Co-Bo built by Metropolitan-Vickers in Stockton-on-Tees, either by looks or performance. Although mechanically and electrically sound in general, the Crossley engines failed to perform satisfactorily and resultantly were usually used in pairs operating the London-Glasgow overnight 'Condor' express freight service or covering for failures on the overnight sleeper. By 1961 all had been dispatched to Barrow-in-Furness operating less demanding Cambrian Coast services but eventually all were withdrawn in October 1967. The locomotive No. D5705 seen here outside Kentish Town shed in September 1959 survived for some time thereafter for use by the Research Division. Being dumped outside Swindon station for several years it subsequently found salvation and preservation at the East Lancashire Railway. *(Flint & Harbart)*

Cricklewood shed often felt like a place where time stood still and this 5 June 1937 view could easily have been re-created well into the 1960's. Resident Fowler LMS 4F 0-6-0 No. 4298 didn't quite make it though being withdrawn during week ending 12 December 1959 having spent many years bouncing between Cricklewood and Kentish Town depots. The special working indicator on the buffer beam is interesting as, being a Saturday, no doubt this predominantly freight-designed locomotive has been pressed into excursion use and following coaling will likely be heading into St. Pancras to head one of the many summer excursions which passed the author's home near South Tottenham en route to Southend, Clacton or other nearby resorts. Jinty 0-6-0T No. 7432 is another locomotive which was still to be found in North London during the early Sixties despite transfer away for several years to the Birmingham, Sheffield and Liverpool areas. Withdrawal from Willesden during week ending 21 August 1965 more or less coincided with the shed's final closure. *(Transport Treasury)*

Another view of Cricklewood shed yard (14A) shows two of the real stars of the Midland main line, at least in respect of freight. Coal was still very much king in the 1920s and regular heavy trains ploughed their way south en route from the Nottingham coalfields to London which required uneconomic double-heading from Toton down to Brent. In 1927 Fowler designed and commissioned 33 Garratt 2-6-0 & 0-6-2T locomotives to dominate these trains for the next 25 years. Unpopular with the crews, especially when running tender-first, these engines were alas heavy on coal consumption but overall they served their purpose whilst not wearing too well, partly due to the wartime conditions which prevailed. No. 7975 (later BR No. 47975) had a working life of 24 years and 10 months being withdrawn in July 1955. This narrows down the date of this image to the early fifties as Toton shedmate alongside was not re-numbered to 48334 until August 1950. *(Flint & Harbart)*

This delightful shot of Gresley Class A3 Pacific No. 59 waiting patiently for the road at the platform end of Kings Cross station has so little in common with any present day comparison of the ex-Great Northern Railway terminus. All buildings on the horizon have certainly disappeared along with the signal box, lighting column, Gresley wooden-bodied coach and so much more. A bit of detective work limits the date to between 26 October 1946 and 8 July 1948 when the number 59 was carried. The loco, located at this time to the nearby Top Shed, was named *Tracery* on 25 March 1925 after the winner of the 1912 St. Leger horse race, a common trend with several other class members. Carrying the LNER apple green livery on a brown chassis and frame, there appears to be a reasonable amount of coal in the tender suggesting it relieved another locomotive at either Grantham or Peterborough and will now retire to Kings Cross loco for turning. *(Flint & Harbart)*

Reversing down onto a northbound departure on the same day is one of the Gresley's most famous masterpieces, the A4 Pacific No. 15 (later BR No. 60015) *Quicksilver*. This was one of four locomotives chosen by the LNER publicity team to cash in on King George V's Silver Jubilee in 1935 by introducing a new high speed service between Kings Cross and Newcastle using dedicated silver painted rolling stock. The 7-coach prestige train comprising two articulated twin sets and one triple set was later supplemented by an eighth coach and ran until the outbreak of World War II in 1939. The Grantham-based locomotive is clearly reversing down onto a northbound departure, however a question arises over the livery. It is reported as carrying wartime unlined black until 1947 when it was then alleged to have received garter blue on black frames. Neither appear to be borne out by this photograph where either light blue or silver seem more likely. *(Flint & Harbart)*

The same location a few years later on 9 August 1953 with Grantham-based Peppercorn 'Pacific' No. 60148 *Aboyeur* getting the road (M1), presumably heading to Kings Cross shed for re-fuelling and servicing. This tragically short-lived locomotive is just six years old when seen here but would be withdrawn from Gateshead shed on 21 June 1965 and sold for scrap two months later. Of additional interest here is the condensing Class N2/2 0-6-2T stabled at the tunnel-mouth head shunt awaiting its next empty coaching stock duty, whilst creeping past York Road platform and heading for the Widened Lines is a rake of vans with trailing guards van. Following the LNER tradition of naming locomotives after race horses, it seems *Aboyeur* was a British thoroughbred and sire but only ran seven races and winning twice. A French word meaning heckler or barker, a new-build Class A1 Pacific still graces our network with a quite distinctive 'bark', No. 60163 *Tornado*, all 49 of the original class being unceremoniously disposed of by the summer of 1966. *(Flint & Harbart)*

Edward Thompson's LNER Class A2/2 were never the most attractive of Pacific designs but much has to do with making something out of a bad job. Taking office as Chief Mechanical Engineer on the death of Sir Nigel Gresley in 1941, Thompson was left with six seriously flawed Class P2 2-8-2 express passenger locomotives and had little choice under wartime restrictions to re-build these as Pacifics under the Class A2/2 label. Adhesion problems initially restricted their intended use in Scotland whereas frame movement, vibration and loose fittings meant the whole class spent considerable time in Works with only one exceeding a million miles in traffic. Their general ungainly appearance and unattractive small smoke deflectors did little to help their appeal and No. 60505 *Thane of Fife*, seen here on Kings Cross Loco in September 1958, was the first to be withdrawn on 10 November 1959 following its re-build in 1943. The run down area surrounding Kings Cross and St Pancras has changed significantly since the latter was re-built to accommodate Eurostar high speed services and everything behind the locomotive has disappeared without trace apart from three gas holder frames. Removed and stored during new railway construction works, they have now been re-erected in the vicinity and substantially modified to frame new luxury accommodation overlooking the revitalised canalside community. *(Flint & Harbart)*

Although Hill's GER Class N7 is mostly associated with the 'Jazz' trains operating out of Liverpool Street or on rural lines in Essex or East Anglia, a batch operated the suburban lines out of Kings Cross for a while in the 1950s and early 1960s. Here we find No. 69632, an N7/5 version of the 3MT-rated 0-6-2 Tank locomotive, with a fine head of steam awaiting departure from the suburban platform with what appears to be semi-fast service in the autumn of 1958. Allocated to Top Shed (34A) it would be returned to more familiar territory at Stratford in April 1961 before withdrawal in September 1961, shortly before the end of steam at Liverpool Street. These platforms are not to be confused with the steeply graded and severe Hotel Curve from which trains on the Down Widened Lines from Moorgate and Blackfriars emerged. Platform 14 (previously 16) is hidden in this view beyond and beneath the rake of stabled coaching stock. *(Flint & Harbart)*

A very familiar view to ardent enthusiasts in 1959 would have been the arrival of through trains from Cleethorpes, probably loaded with returning holidaymakers. The carriage nameboards suggest this is a scheduled service and Immingham-based Thompson LNER Class B1 5MT 4-6-0 No. 61142 looks reasonably tidy which wouldn't last much longer with its imminent transfer in December to Colwick. Near identically numbered sister locomotive No. 61042 was a tediously regular performer on the 'Cambridge Buffet Express' with its white smokebox door hinges and painted buffer beam and was often received very unfairly with jeers by spotters hoping for a little more variety. Lurking in Kings Cross loco yard to the left can be seen D5308, a Birmingham Railway Carriage & Wagon Works Type 2 diesel electric Bo-Bo, 20 of which were delivered new to Hornsey from late 1958. Later BR Class 26, a further similar batch were delivered to Cricklewood before both types migrated north to Scotland where for a non-standard type they proved quite successful compared with other less respected classes. *(Flint & Harbart)*

During busy spells at any London terminus, a re-fuelled and serviced locomotive could be pressed into service to haul empty coaching stock into the station prior to its own departure. Top Shed-based Class B1 4-6-0 No. 61364 is seen buffered up tender first to the stops sometime in 1960/61 awaiting its release but will no doubt give assistance to the departing service through the humid and greasy Gasworks and Copenhagen Tunnels. In June 1962 it would be relegated to New England shed to extend its short career of less than 12½ years by 3 months. Edward Thompson's 410-mixed traffic Class 5MT locomotives should be considered as a great success bearing in mind construction commenced under wartime and economic restrictions, generally popular with crews throughout the LNER apart maybe from the occasional rough riding. Fortunately two examples have been preserved, No. 61264 on the Great Central Railway (Nottingham), the only ex-LNER salvaged from a scrapyard, and No. 61306 *Mayflower*, a main line approved locomotive which carries the name of withdrawn sister No. 61379. *(Barry Richardson)*

Don't try this at home! Long gone are the days of being invited onto the footplate of a moving steam locomotive on the main or subsidiary lines but our photographer was either officially employed or had influence back in the days of this late 1962 view. Gresley GN Class A3 Pacific No. 60046 *Diamond Jubilee* was on a 7 month re-allocation to New England before return to Grantham prior to final withdrawal on 16 June 1963 but was clearly still employed on top link work. The Up express working will be steadily braking as it is about to enter the sulphurous Gasworks Tunnel on the final approach to Kings Cross. The only person allowed to look out of a carriage window these days, even if it did have any that opened, is the guard as common sense seems to have been superseded by Health and Safety Regulations! *(Barry Richardson)*

The imposing shape of an English Electric 'Deltic' Type 5 at rest on arrival at Kings Cross is a sight to behold even to a fervent steam fan. All is not as it seems though in this late 1970s image as Finsbury Park's No. 55010 *The Kings Own Scottish Borderer*, having recently arrived with an Up InterCity express from the north-east, will have observed the presence alongside of its ultimate replacement in the form of the Class 43 InterCity 125 High Speed Train. Despite looking a little work-tired with ingrained staining on its BR blue livery, No. 55010 soldiered on until the end of December 1981 before it was broken up at Doncaster Works the following May. *(Arthur Turner)*

To be fair, one of the great railway successes of the late 20th century was the introduction of the InterCity High Speed 125 diesel unit, improving speed, capacity and comfort on the East Coast main line, Midland main line and Western Region lines to South Wales and the West Country. In its striking livery and under the original numbering system, unit No. 254012 is entering Kings Cross station with a service from the north shortly after the introduction in May 1978 and passing the soon to be redundant Kings Cross Loco servicing facility. *(Arthur Turner)*

Situated between Gasworks and Copenhagen Tunnels, the descriptively named Goods & Mineral Junction signal box forms the backdrop of this late 1950s view of Belle Isle where this GN main line passes under the viaduct and embankment of the North London line. Gresley Class A3 Pacific No. 60055 *Woolwinder*, another St. Leger winner, is still finding its feet departing from Kings Cross with a Down express whilst an unidentified classmate awaits the road into the terminus closely followed by a Brush Type 2 A1A-A1A and vans. The rather inappropriately named Belle Isle was the junction for access to Top Shed, having been the location in the 19th century of refuse yards, chemical works, manure factories and a 'knackers yard' to name but a few. *(R C Riley)*

Sir Nigel Gresley's Class V2 7P6F 2-6-2 is often described as the locomotive that won the war, reliably keeping the main freight route along the ex-Great Northern Railway running under extremely adverse conditions. The doyen of the class No 60800 *Green Arrow* is captured outside the running shed at Kings Cross, its home depot in the late 1950s alongside shed-mate Class A4 No. 60014 *Silver Link*, both being prepared for their next main line duties. *(R C Riley)*

Regular steam working at Kings Cross ended with the beginning of the Summer 1963 timetable but steam still covered diesel failures for some time thereafter. Accordingly Top Shed (34A) closed in June with demolition following shortly after. Suburban services through the Widened Lines were by now totally in the hands of DMUs or diesel-hauled rolling stock with officially some of the stud of Class N2 0-6-2T transferred during week ending 26 May 1962 to New England shed.  Well away from the running sheds a line up of both N2/2 and N2/4 sub-classes are observed. With chimneys sacked over but otherwise seemingly intact., Nos. 69523, 69568, 69531, 69593 and 69538 on paper made the move north to Peterborough but it is hard to envisage any of the condensing apparatus-fitted tanks finding any employment before they were all officially withdrawn in the September of 1962. *(Barry Richardson)*

Another late 1970s view spotlights one of the recently introduced dual-voltage Class 313 EMU suburban 3-car sets, No. 313010 awaiting departure from Finsbury Park to Kings Cross. Classified as second generation units, they were based on the BREL 1971 prototype design. The Great Northern fleet survived until 2019 when they were completely replaced by the current Class 717 units although a few cascaded onto the Southern's Coastway services and hung around until mid 2023, hardly popular though with their lack of air-conditioning and toilet facilities. When loco-hauled suburban stock still monopolised services in and out of Kings Cross, a fly-over from the ex-Edgware and Alexandra Palace branch connected to the Up main lines and was regularly utilised for empty stock movements and could be clearly viewed from the famous 'spotters platform' in Finsbury Park, just to the north of this spot. *(Arthur Turner)*

At Alexandra Palace terminus station on 15 May 1956, Class N2/2 0-6-2T No. 69506 is seen running round a 'Special' train whilst liberated passengers roamed at will around the abandoned premises. The station was built and opened by the Muswell Hill Railway on 26 May 1873 but two weeks later the palace burned down with subsequent services significantly reduced. However, soon after restoration of the palace, demand failed to materialise so the name Alexandra Park was dreamed up to seek a wider audience. This didn't work so the 'Palace' title was reinstated before the Great Northern Railway assumed control under the Railways Act of 1921, becoming part of the LNER in 1923. The line itself branched off the Finsbury Park to Edgware and High Barnet line which in 1935 London Underground planned to take over, modernising it for use by tube stock and connecting it to the Northern Line but works were interrupted by World War II. Upgrading works on the Highgate to Alexandra Palace line were cancelled, services were further reduced and with ever-decreasing patronage this section closed to passengers on 3 July 1954. Freight continued until 18 May 1957 and the tracks were still used between Finsbury Park and Highgate for transfer of LT stock to and from the Highgate depot, albeit hauled by battery locomotives as the line was never electrified. Much of the trackbed from the palace now forms part of a walk through woodland and offers views over the suburbs from an elevated position. *(A E Bennett)*

A little to the north of Finsbury Park is Hornsey and the vast freight yards of Ferme Park. Hornsey MPD was an eight road straight shed constructed by the GNR in 1899, superseding earlier smaller two-road sheds on this site and nearby Wood Green. The depot provided a host of small shunting locomotives to serve Ferme Park Yard and for cross-London freight trains routed south via Snow Hill Tunnel and Blackfriars. The shed's complement of N2 and L1 tanks also provided power for the extensive suburban services operating from Kings Cross and Moorgate via the Widened Lines as well as empty stock workings. Locomotives from further north arriving with heavy freight were usually serviced here but on a very cold-looking 9 March 1958, resident Gresley Class J6 GN 2P3F 0-6-0 No. 64233 is heading off shed. Any diehard enthusiast observing matters from the station's 'grandstand' footbridge should have been able to make out the identity of a couple of visiting WD 'Austerity' 2-8-0s stabled outside the shed. *(David Moore)*

Turning through 90 degrees and looking north from the same location on the footbridge on 14 July 2019, a pair of Class 365 'Networker' EMUs are pulling away from the modern Hornsey Traction Maintenance Depot in readiness for the evening's rush hour. However, their days were numbered and by 2021 Great Northern had withdrawn the last of these units introduced in 1994, in favour of new Class 387 'Electrostar' units built by Bombardier. In the days of steam and diesel hauled suburban services, the running lines to the left were protected by a very impressive 10-signal gantry with a smaller 4-signal gantry covering the Up Slow and engine shed access lines. *(Author)*

Suitably buffed up and polished outside Broad Street station on 5 May 1956 is North London Railway Class 75 0-6-0T No. 58859 built in 1889 at Bow Works to an 1879 design by J. C. Park. These powerful but compact engines were designed for shunting the London Docks but transferred to the LMS on grouping. Originally numbered 7522, it was employed for some years in Birkenhead Docks. One of 14 surviving examples taken on by British Railways, this locomotive returned to Devons Road, Bow and continued work until October 1957. This day is probably the highlight of its career as it was rostered to head the Broad Street-Poplar leg of the LCGB's 'Poplar & Edgware' rail tour to Poplar, in the run down heart of the docklands. Out of interest, the 3rd Class fare for this afternoon tour was 7/6d or 37.5p in modern parlance. The locomotive in the background is No. 43020, one of Devons Road's stud of Ivatt Class 4MT 2-6-0s, which is shunting Broad Street Goods. Both passenger and goods stations were situated on arches, high above the adjacent Liverpool Street station which meant goods had to be transferred to ground level via wagon turntables, transverse lines worked by capstans and three No. 25 ton and two No. 20 ton hoists. Locomotives were barred from entering the goods shed due to fire risk which led to a very labour intensive operation. However, traffic levels were high, including merchandise and perishable goods, most notably fish from Fleetwood. *(A E Bennett)*

A typical quiet, virtually deserted platform end scene at Broad Street outside rush hour is recorded towards the end of the 1950s. The 3-car open set EMU on the left was part of a batch built by Metropolitan-Cammell for the LNWR from 1914, incorporating the Swiss manufacturer's Oerlikon traction motors with both third and fourth rail current collectors. As well as operating over the Watford DC lines, these units also ran on shared tracks with London Underground District and Bakerloo Line services. The train on the right is a member of BR Class 501, 3-car units introduced between 1957 and 1958 thus dating the image to between 1957 and 1960 when the last of the Oerlikon units were withdrawn. These trains monopolised the Euston-Watford and Broad Street-Richmond/Watford services until 1985. Fitted with barred windows, due particularly to limited clearances within Hampstead Tunnel, these trains based on the Southern Region EPB stock, earned themselves the nickname of 'Jail Units'. Closed on 30 June 1986, all remaining services were diverted into Liverpool Street via a new spur off the North London Line whilst the modern Broadgate office and shopping complex has removed any trace of the historic North London Railway terminus. *(A E Bennett)*

The state of dereliction is well advanced as participants of the 5 May 1956 'Poplar & Edgware' rail tour spill out over the tracks at Poplar Docks where, beside the elevated Bank signal box, is something of a rarity, LMS prototype battery electric locomotive B.E.L. No. 1 formerly numbered 1550, a machine which worked at the coal depot near West India dock between 1913 and 1964. As there was no rail access, a hydraulic hoist was used to get wagons onto site, thus limiting the size, height and weight of the locomotive which had to use the hoist for maintenance, overhaul etc. *(A E Bennett)*

The excursion returned from Poplar via East Ham and Stratford behind Plaistow-based Jinty 0-6-0T No. 47484 as far as Canonbury where it handed over to a Kings Cross loco, Gresley Class N2/2 0-6-2T No. 69506, recorded earlier at Alexandra Palace, still retaining its condensing apparatus and small chimney for working through the Metropolitan-gauged tunnels to Moorgate. Canonbury station was opened by the North London Railway just east of the current location under the name of Newington Road and Balls Pond Road. Passenger services operated to and from Broad Street between 1875 and 1976 when a new surface connection between Finsbury Park and Drayton Park enabled through running to Moorgate. The route is still an important freight corridor whilst many changes have subsequently occurred with new signalling and platform extensions as part of the London Overground works. The station is now served by both the Mildmay Line to Stratford and Richmond/Clapham Junction and the Windrush Line to West Croydon or again Clapham Junction but in the clockwise direction. *(A E Bennett)*

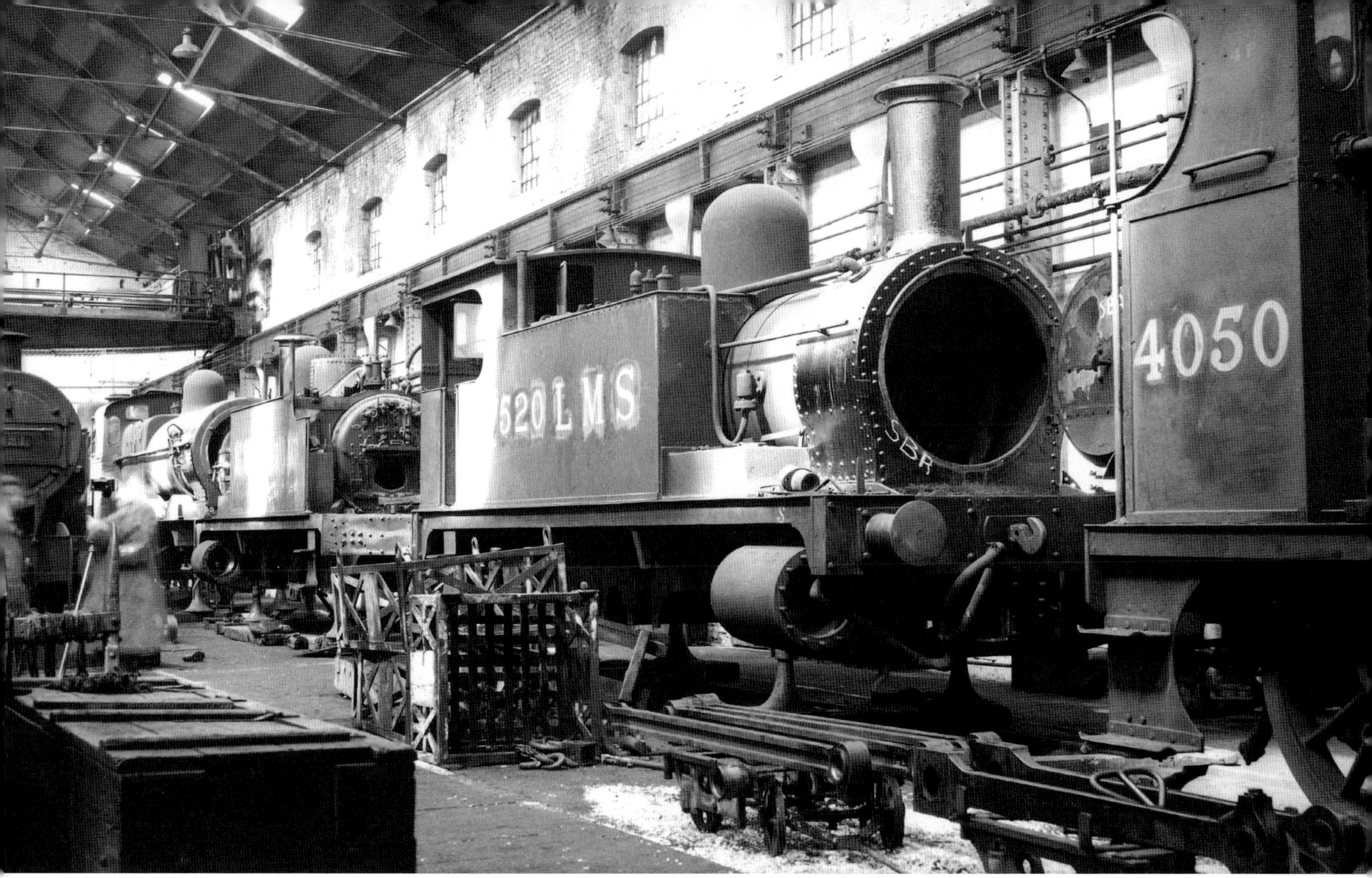


Bow Railway Works was built by the North London Railway on a 10 acre site in 1853 to maintain and repair both locomotives and wagons. From 1860 it was enlarged to facilitate locomotive construction, the first being a 4-4-0T No. 43 which incorporated a specialised bogie for stability at speed, designed by locomotive superintendent William Adams. Adams was succeeded in 1882 by J. C. Park and construction of a new erecting shop led to the building of a series of 4-4-0 and 0-6-0 tank engines until 1901, 14 of which survived into British Railway days. Construction ceased in 1908 when the NLR merged with the London & North Western Railway but repairs continued despite heavy damage by the Luftwaffe. Under the banner of the London Midland & Scottish Railway from January 1923, the Works assumed responsibility for the repair of London, Tilbury & Southend Railway rolling stock following closure of Plaistow Works in 1927. Having led the development of the Automatic Warning System and extended its scope to include repairs of diesel locomotives from nearby Devon Road MPD, Bow Works finally closed in 1960 when all work was transferred to Derby Works. This 1946-47 interior shot shows various locomotives under repair, from right to left identified as Fowler MR 4F 0-6-0 No. 4050 (BR No. 44050), Park designed in-house 2F 0-6-0Ts Nos. 27520 (BR No. 58858) and 27513 (BR No. 58854), an unidentified MR 4F 0-6-0 and Whitelegg LTSR 3F 4-4-2T No. 2121 (BR No. 41939). *(Transport Treasury)*

In this 1960 view of Liverpool Street station, Hill GER Class N7/5 0-6-2T No. 69663 accelerates away towards Worship Street overbridge with a Down 'Jazz' commuter service towards either Chingford or Enfield Town or possibly Palace Gates. These Stratford-based locomotives served the suburban lines admirably until electrification of the suburban lines made them redundant. There is no sign here of any overhead catenary being erected at this time and the rapid turn round at the terminus was a hallmark of these reliable locomotives as well as their Class F4 2-4-2T predecessors. The original trains introduced in 1920 comprised of 16 four-wheel coaches with stripes above the windows highlighting the class of travel, yellow for First, blue for Second and unmarked for Third. Additionally each route of travel was identified by coloured boards such as pink for Palace Gates via South Tottenham hence the nickname of 'Jazz trains' which they retained until the end of steam. Comfort was clearly second in priority to capacity and narrow 6-aside compartments were retained in the Gresley timber-bodied 'Quint-art' 5-car sets which had adjacent coaches sharing a single bogie. Gresley adopted a similar design for his 'Quad-art' stock which operated on the Great Northern lines into Kings Cross and Moorgate. Towering above the pleasingly clean locomotive is the substantial structure of Skinner Street Junction's 84-lever signal box built in 1892. *(R C Riley)*

During the late 1950s and into 1960 Stratford shed prided itself in turning out two pristine locomotives to serve as West and East Side Pilots at Liverpool Street station. This 1960 view captures Holden GER Class J69/1 0-6-0T No. 68619 in immaculate pseudo GER blue lined-out livery with BR decals, in stark contrast to much of the other motive power and rolling stock, especially that employed on the 'Jazz' suburban services. Its partner in crime was usually the similarly painted Class N7 No. 69614 whilst the custom was reintroduced briefly in the 1980s when Class 08 DM shunter No. 08833 appeared in the same livery adorned with the wording 'Liverpool Street Pilot' and the famous cockney sparrow cartoon. More camera shy in the background is DE No. D0226, one of two prototype 0-6-0 shunters built by English Electric in 1956, withdrawn in 1960 but now preserved on the Keighley & Worth Valley Railway. D0227 was virtually identical but with hydraulic transmission, withdrawn in 1959 and returned to Robert Stephenson and Hawthorns, part of the group. *(R C Riley)*

Gresley's 'Sandringham' or 'Footballer' Class B17 4-6-0s built for operation on the Great Eastern main line were regularly seen passing through Stratford in London's East End. Ipswich-based B17/6 variation No. 61618 *Wynard Hall* entered service on 5 November 1930 as a B17/2 and, having been fitted with a B1-type boiler, had spells at both Stratford and Cambridge from where it was withdrawn on 23 January 1960. Heading a Down parcels train in the late 1950s its days were clearly numbered as any top link work had by now been taken over by the new BR Standard Pacifics. Wynard Hall is a grand country house situated near Stockton-on-Tees in County Durham and it is very unlikely the locomotive ever passed that way other than maybe on visits to/from Darlington Works. *(Nick Nicholson)*

Storming up the bank near Bethnal Green on the Great Eastern main line in the late 1950s is one of Stratford's 'Footballer' Class B17/6 4-6-0 No. 61654 *Sunderland*. In 1937 two of the class were cosmetically streamlined for publicity with no significant performance improvement and ultimately in 1951 *East Anglian* and *City of London* were stripped of their cladding. The choice of football clubs represented by the naming never tied in geographically with the areas in which the locomotives operated, Ipswich Town and Colchester United probably feeling the most peeved. These locomotives were phased out between 1952 and 1960 by Thompson's reliable Class B1 4-6-0s and the powerful BR Britannia Pacifics, much more suitable for express services to Ipswich and Norwich. Curiously the name of Sunderland Football Club lives on in the guise of GB Railfreight's modern Class 66 No. 66725 with the clubs of Derby County and Huddersfield Town being similarly celebrated whilst Sheffield United have now been deposed by arch rivals Sheffield Wednesday! *(R C Riley)*

0-6-0 tender locomotives were always abundant in the vast expanse of Stratford motive power depot and scarcely turned too many enthusiasts' heads at the time. Prominent amidst the piles of steaming ash in the yard in the late 1950s is a resident Hill ex-GER 5F Class J20/1 No. 64677, built in December 1920 and probably rarely moving too far away from Stratford during its career of just under 41 years. Gresley 4P5F J39 No. 64784 behind is a member of his most numerous class, 289 in total being built over a period of fifteen years. During the author's formative years these were not so common in East London and No. 64784 somehow escaped the attention of his Bic pen and ruler. In its service life of 31 years and 2 months it travelled widely and was based at York, Glasgow Eastfield, Edinburgh St.Margarets, Carlisle Canal, Thornton Junction, Dundee Tay Bridge and Norwich Thorpe but during its final 10 years at Stratford it may have been out-based at any of the depot's many sub-sheds. *(R C Riley)*

James Holden's GER Class D56 (classified D16 by the LNER) were unquestionably his most famed achievement, an inside cylinder 4-4-0 locomotive capable of hauling the fastest express services and widely acclaimed as one of the most elegant designs of the pre-grouping period particularly when painted in the company's blue livery. Nicknamed 'Claud Hamiltons' after the plates carried by the pioneer of the class in recognition of the company's chairman, these were the flagship locomotives on the main line during the Edwardian period. Entering service in December 1909 as No. 1829, BR No. 62580 underwent two major modifications in 1930 and 1948 whence it subsequently assumed the numbers 8829, 7699 (not carried) and 2580 before being withdrawn by British Railways from Kings Lynn shed on 9 June 1958. Sadly it was sent to Stratford Works for breaking where we find it, carrying its 31C shedplate and awaiting the cutters torch the following month. *(R C Riley)*

Stratford Works and locomotive sheds covered a vast area and searching for any of the various Departmental Locomotives could be like looking for the proverbial needle in the haystack. Dick Riley succeeded however and found No. 33 in steam and working hard in early 1961. A Hill GER Class 74 0-4-0T, it was at Stratford in 1921 as No. 210 subsequently becoming Nos. 7210 and 8129 before entering departmental use in September 1952. Little is known of previous employment but it probably never wandered far from its place of build. Here it seems to be very active with coal piled high on its left hand tank and on top of the firebox as no rear bunker was provided. Withdrawal came during December 1963 and it was later discovered during Spring 1964 in King's Scrapyard at Brunswick Junction, Poplar where it was presumably dismantled in due course. Following the end of steam the Works gradually diminished and finally closed in 1991, all traces now lost beneath Stratford International station and the Westfield Shopping Centre. *(R C Riley)*

**Left:** Another of Stratford's huge allocation of locomotives is Thompson LNER Class L1 4MT 2-6-4T No. 67709, on 20 July 1961 passing the substantial hump marshalling yard at Temple Mills, opened in 1958 as part of the 1955 Modernisation Plan. The residual diesel repair depot alongside, although coded TD, appears never to have had its own allocation of locomotives and closed in 1991 in line with the loss of business in the yard. Nearby nowadays but not on the same footprint, is the Eurostar depot, coded TMI. Around the 2-road covered building are early Brush Type 2 (BR Class 30) A1A-A1A diesels and an unidentified steam 0-6-0 intruder whilst the focus is on No. 67709 heading a North Woolwich to Palace Gates (and Wood Green) local, an infrequent service routed via Stratford Low Level, South Tottenham and Seven Sisters. The section from Seven Sisters to Palace Gates unsurprisingly closed to passengers in January 1963 with the Wood Green vicinity well served by Underground services whilst the eastern end of the branch is now served by the Docklands Light Railway and the Elizabeth Line, North Woolwich closing in 2006. *(David Moore)*

**Right:** Approaching Lea Bridge with a lengthy Up transfer freight from the Lea Valley direction, again on 20 July 1961, is Stratford-based Worsdell GER Class J15 1P2F 0-6-0 No. 65462. Possibly heading for the nearby Temple Mills marshalling yard, another likely destination was the modern hump yard at Ripple Lane in East London opened in 1958 but subsequently closed in 1965 as bulk trains took over freight operation. Lea Bridge Gas Works behind the line of mineral wagons was built by the South Essex Gaslight and Coke Co. in 1852 with the gasometers only finally being taken out of action in 2012 .The site was subsequently re-developed for housing, public open space and riverside amenities, far removed from the image depicted here. Partly hidden behind the empty wagons appears to be one of Stratford's many diesel shunting locomotives in the form of a 204bhp Class 03 or 04 DM 0-6-0. J15 No. 65462 had arrived at Stratford this year from Norwich Thorpe but would only survive until 16 September 1962. *(David Moore)*

65462
959
BOARD

A sharp north to east spur connects Seven Sisters with South Tottenham, a short walking distance away, the route taken originally by Palace Gates to North Woolwich services. South Tottenham is on the original Tottenham & Hampstead Junction Railway and Fowler and Stanier LMS 2-6-2T locomotives used to operate stopping trains between Barking and St. Pancras via a now closed spur at Kentish Town. Handing over to DMUs the trains were later diverted to Gospel Oak and on 2 June 2016, BR Bombardier Class 172 'Turbostar' No. 172 004 is about to bear off left at South Tottenham East Junction and head towards Barking with a London Overground acronymic 'GOBLIN' service, on what is now known as the Suffragette Line. At this time the line was not fully electrified, the catenary in place serving diverted services out of Liverpool Street and through freight workings. *(Author)*

Stephen Dewar Holden designed his Class F7 2-4-2T in 1909 to replace the J65 0-6-0T accredited to his father James whom he succeeded as Chief Engineer of the Great Eastern Railway. Only twelve were built and they were not deemed to be a great success with only two surviving into Nationalisation. Their rather incongruously large cabs earned them the nickname of 'Crystal Palace Tanks' and their duties were normally limited to light branch line work. No. 8303 fulfils that role in this 1937 image recorded at Seven Sisters where it has arrived on the curved timber platforms of the branch from Palace Gates (Wood Green) with at least the first coach being a non-corridor teak vehicle with its clerestory roof providing supplementary light and ventilation. Elderly Class 2-4-2T continued to operate on this branch into the 1950s followed by J68 and J69 0-6-0Ts followed by larger Class N7 0-6-2Ts and Class L1 2-6-4Ts although early DMUs were also tried prior to closure to passengers in January 1963. *(Neville Stead Collection)*

Probably the greatest claim to fame for the Palace Gates branch was the hosting in Noel Park's goods yard of a 3-day railway exhibition in September 1958 commemorating the 750th anniversary of Wood Green becoming a chartered borough, reportedly attended by 14,000 people. Star of the show was Gresley A4 Pacific No. 60022 *Mallard* alongside a year old BR Standard Class 9F 2-10-0 No. 92196, a Saltley engine which had a wastefully short career of just over 10 years. Other steam locomotives were ex-GNR Class C12 4-4-2T No. 67352, suitably re-painted and lined out by apprentices in Doncaster Works and immediately condemned after the exhibition. Faring much better was Ivatt GN 0-6-0ST Class J52 No. 68846 which was shortly in 1959 to become the first locomotive to be sold to a private owner, Captain W.G. Smith and permitted to operate charter trains on the network. *(Roy Vincent)*

New diesels were also on display in the shape of English Electric Type 4 Co-Co No. D208 (later BR No. 40008), Birmingham RC&W Type 2 Bo-Bo No. D5300 (later BR No. 26007), Class 10 0-6-0 DE shunter No. D3490 and a brand new Craven 2-car DMU. The highlight however for an eleven year old school boy, who still has the programme in his hoard at home, was undoubtedly spending 2d in old money to be suspended in an open wagon body hoisted high above Gresley's record-breaking Pacific by a 1939 steam crane. *(R C Riley)*

Industrial steam could still be seen until 1970 in the Lea Valley from the electrified main line between Tottenham Hale and Cheshunt shunting coal wagons at Rye House Power Station. Built by Robert Stephenson & Hawthorn in 1949, 0-6-0T No. 7597 spent its entire working career at the CEGB establishment until it was purchased privately by the 7597 Fund and after several years at Chappel & Wakes Colne (East Anglia Railway Museum) eventually moved on to its present location, the Lincolnshire Wolds Railway, where it currently awaits a boiler overhaul. *(Neville Stead Collection)*

The GER branch to Ongar opened as an extension of the line to Loughton in 1865. Under the LPTB's New Works Scheme in 1949 the Central Line was extended to Epping but opted not to electrify the single line to Ongar on economic grounds. Thence for a period of eight years British Railways operated the line on hire to London Transport and this is the case seen here in the early 1950s with Stratford's 1905-built Class F5 2-4-2T No. 67207, a Holden re-build of a Worsdell design, arriving with a service from Ongar. Known as 'Gobblers' due to their insatiable appetite for coal, the class were eliminated by 1958. Alongside on the adjacent platform is one of the Central Line's Standard Stock 4-car units, Driving Motor No. 7417 being identifiable. Ultimately electrification of the branch was completed in 1957 on a strict budget meaning only 4-car trains could run through to Ongar. Inevitably as the anticipated patronage never materialised, the line was closed completely in 1994 after stealth cut-backs. Happily the line still exists operated as a heritage-line by the Epping & Ongar Railway but with no through running possible. *(R C Riley)*

The eastern terminus of the GER branch from Palace Gates was North Woolwich in the heart of London's docklands. Originally the terminus of the Eastern Counties & Thames Junction Railway from Stratford, it opened in 1914. A Stratford-based engine from January 1960 to its withdrawal in September 1961, Hill GER Class N7 3MT No. 69724 is running round its train with the adjacent goods yard appearing to be very much still in business but succumbing at the end of 1970. From 1963 diesel traction took over this route which was cut back to Stratford with a rush-hour service to Tottenham (re-named Tottenham Hale in 1968) to coincide with the opening of the LT Victoria Line. In 1979 the route became an extension of the North London Line and was electrified under the third rail system to extend the service round North London to terminate at Richmond. Temporary closure of the line ensued from May 1994 to October 1995 during construction of the Jubilee Line but closed again in 2006, the area deemed to be better served by the DLR King George V station. From 1984 to 2008 the Grade II listed original station served as a museum dedicated to the Great Eastern Railway, the collection now dispersed. *(R C Riley)*

Another Class N7 0-6-2T No. 69670 is later captured approaching North Woolwich from Silvertown, running parallel to Albert Road past one of the many bomb sites which prevailed following the last war. The gallery of cranes lining Royal Albert and King George V Docks tower over the residential properties interspersed frequently with public houses, whilst the red and black chimney appears to indicate the presence of a Cunard Line ocean-going vessel. This branch line in its heyday was heavily used by dockers and ancilliary workers whilst others would make their way to the free ferry adjacent to the terminus, crossing the River Thames and making their way to Woolwich Arsenal, both major centres of employment during both World Wars. *(R C Riley)*

Probably due to the vast range of main line railway interest in North London, the industrial scene was largely ignored by enthusiasts and photographers. Fortunately with the threatening elimination of steam, some took the time to seek out the hitherto unknown and on 16 March 1957, the RCTS organised a tour which included the Prince Regent Tar Works in the highly industrial area of Silvertown in the heart of the docklands. In operation was a Bagnall 0-4-0, builders No. 2851 built in 1947, a fireless locomotive and essential in such a combustible and noxious environment, particularly as the company also operated as timber merchants in the same vicinity when the tar would have been used as a preservative. Ironically, the parent company Burt, Boulton and Hayward Ltd. survived a significant fire on the premises in 1937 which is believed to have started in the store which maybe led to the requirement for such a locomotive. *(R C Riley)*

Probably the least known of London's termini north of the river is Fenchurch Street, at the City end of the London, Tilbury & Southend Railway and the London & Blackwall Railway. With just four platforms and no adjoining London Underground station, the operation of services to this day is very intensive but to a railway enthusiast rather predictable, particularly with the rapid turn-around of the current Class 357 electric multiple units. Prior to electrification in 1961, services were in the hands of 'modern' BR Standard Class 4 2-6-4T locomotives, one of which, No. 80100, is seen here approaching the terminus on 24 March 1956 with the 2.45pm from Shoeburyness. East London is still showing signs of damage inflicted by the Luftwaffe during the war and the far from new non-corridor rolling stock did little to appease commuters who regularly complained about 'the misery line' and the persistent overcrowding and uncleanliness of the trains. *(Leslie Freeman)*

This exterior view of Fenchurch Street station graphically shows its confined location and uninviting approach on an evidently wet and depressing 21 January in 1961. The mandatory tobacconist outlet seems wedged between the station and the adjacent business premises of Lilley & Reynolds Ltd, a firm of compass makers and adjusters, trading from 14 London Street since 1854. The vintage car seems rather incongruous parked behind the Lambretta scooter whilst a taxi is coyly hidden behind a section of unkempt hoarding. The British Railways delivery van is reasonably modern for the era but overall an air of despondency prevails. Plans have previously been mooted to move the station eastwards and expand the platform capacity to six, built partly on the site of the Docklands Light Railway Tower Gateway station which would ultimately close. Like other plans to incorporate the terminus was the proposed Fleet Line (now the Jubilee Line) but this has sadly not materialised to date. *(Leslie Freeman)*

Scattered around North London were many uncelebrated goods yards and warehouses, all requiring the service of shunting engines. Situated alongside the River Lea in East London was Regents Wharf, home of the Regent Oil Company, once a common brand name on petrol pumps throughout the capital. Seen on 28 September 1956, Andrew Barclay DM 0-4-0 shunter No. 11503 was part of a small group of four short wheel-based locomotives maintained at Stratford and later to become BR Class 01 and re-numbered D2953 in 1960. After 10 years in service it was sold on to Thames Matex Ltd in 1966 and later preserved, now residing at Peak Rail in Derbyshire. *(Transport Treasury)*

Plaistow MPD, a London, Tilbury & Southend Railway shed providing motive power for services in and out of Fenchurch Street, was situated to the south of the main line between West Ham and Plaistow stations. A large shed handling mainly small engines, it officially closed to steam on 2 November 1959 and completely on 18 June 1962 but this winter 1960/61 shot clearly shows Jinty 0-6-0T No. 47328 in steam acting as shed pilot and still carrying the appropriate 33A shedplate. Contemporary reports indicate its transfer to Tilbury shed following the official closure whilst Shoeburyness-based Fairburn 2-6-4T No. 42224 also appears unperturbed amongst the piles of recently discarded ash. Plaistow was also home to a locomotive works but this closed in 1925 with work being transferred to Bow. 1960's musician Joe Brown worked in the shed for two years, eventually making the grade of fireman before moving on to greater things, whilst his family ran the local 'Sultan' public house, demolished in 2005. The site of the shed is now the home of the East London Rugby Football Club. *(Barry Richardson)*

Ripple Lane Traction Maintenance Depot, coded RL, was similar to the facility at Temple Mills, both sub-sheds of Stratford and therefore never officially allocated any motive power although three No. Class 08 shunters were normally out-stationed either here or in the nearby Ford Works. Situated near the marshalling yard between the LT&S Up and Down main running lines from Barking to Dagenham Dock, it comprised a 4-road open ended concrete and glass roofed building with each road having an inspection pit. In this 1980s view the complement of locomotives includes a nominal Class 08, five No. Class 47 Type 4s and a pair of Class 37 Type 3s including No. 37706 doing its best to obscure the scene with a no doubt raucous emission of toxic fumes. This locomotive delivered new to Stratford as No. D6716 in June 1961 is still in operation for West Coast Railways whilst the identifiable Class 47 No. 47381 lasted less than 27 years, starting life as D1900 in September 1965. *(Arthur Turner)*

The new Eastern terminus of the Suffragette Line is Barking Riverside serving a regeneration area built over the site of Barking Power Station, part of a stringent planning requirement to ultimately construct around 10,800 homes. Just under a mile of new track mainly on viaduct branches off the LT&S line near the site of Ripple Lane Yard and now provides an interchange with ferry services operated by Thames Clippers from central London. On 2 January 2024 at this currently rather desolate location, BR Class 710 Aventra EMU No. 710259 awaits departure from the island platform with a service to Gospel Oak, the line fully electrified throughout since the summer of 2022. *(Author)*